Life Beyond in Close Reach

Discovering Life in our Solar System

Michael A. Koenig

To My Family

Index

Introduction

Life, something so common in our human species everyday lives, from the person walking by to the birds flying above even the smallest of plants are life. The most asked question in the history of our Homo sapiens lives is are there forms of life out there in our galaxy to be even more specific our solar system. This question at first was just a question now with a numerous amount of scientist who have researched and studied for this very topic in result there have been many candidates of extraterrestrial body's that can or is withstanding life. There has been much research done to try and discover life in our solar system for example the many research crafts in space currently discovering things that can help with the push to finding life.

There are future attempts planned to research these possible habitats.

The most likely body's for the occurrence of life are body's that are accessible to the human species. This will cover the eight body's most likely a home for life it's out there and not to far off.

Chapter 2: Mars

Planet makeup

Avg. Distance from Sun- 227,936,640 km.

Radius- 3,397 km.

Mass- 0.107 x Earth

Gravity- 0.4 g.

Length of day- 25 hours

Length of Year- 2 Earth years

Surface temperature- -5 c -85 c

Main Atmospheric Composition- Argon, Carbon dioxide,

Nitrogen

Moons- Phobos and Deimos

Mars is probably the most known planet not including ares. Why is that well the reason being it is the most researched and the quintessence of what people think about when someone says extraterrestrial life.

The current NASA Curiosity rover has made massive discoveries towards the existence of life out side of our solar system. There has now been very much evidence to conclude running water had ounce been on the red planet. This is a big step in confirming life. Life as we know needs certain things to be created and to survive an important thing is the planet which is forming the life to be in the Hannibal zone. This so called zone is in a spot that is not to close the sun it's orbiting but not to far also that it's to cold. If the planet is in that zone then it is able to withstand life on the surface. Another crucial part the different elements needed to withstand life. Finally a good habitat, atmosphere, water, and luck. Then with these things you have life. The search of life on Mars had also been strong even before the first craft to the planet, there has been numerous different signs of pop culture about Mars and its potential for life. Mars is said to maybe have a layer of water or ice underneath it's surface. Even if after many years of searching for Mars life and nothing comes up there could be a whole ecosystem underneath the crust. There are plans now for new vehicles to explore under the surface. a big theory about not just any life but human life is that it actually started on Mars. This theory is called the panspermia theory that life first existed on Mars then somehow traveled to Earth. There are current workings by a private company Mars One to send up human beings to the Martian planet by 2026. The people sent up to Mars will be there preeminently and there will be no one sent back to earth.

The main focus of any space based exploration company now days is on Mars. Mars is a planet so much regarded as having life that it is sometimes hiding the truth that life could be out there on a different planet or moon even more planet or moon likely too.

Chapter 3: Io

Moon makeup

Avg. Distance from Sun- 778,412.020 km.(Jupiter)

Radius- 1,822 km.

Mass- 9x10 to the 22 power kg.

Gravity- 1.796 m/s/s

Rotating planet/body- Jupiter

Distance coronation to planet- 5th

Size compared to other moons on same orbit- 3rd

Io, a moon very unfamiliar to the normal human being. Most

moons other then Earths are not really regarded as famous or

very well known in the common day world. Io has been said to

resemble a pizza and it sort of does but well there's more to it

then just that, life. Yes, life has been said to be on this moon just

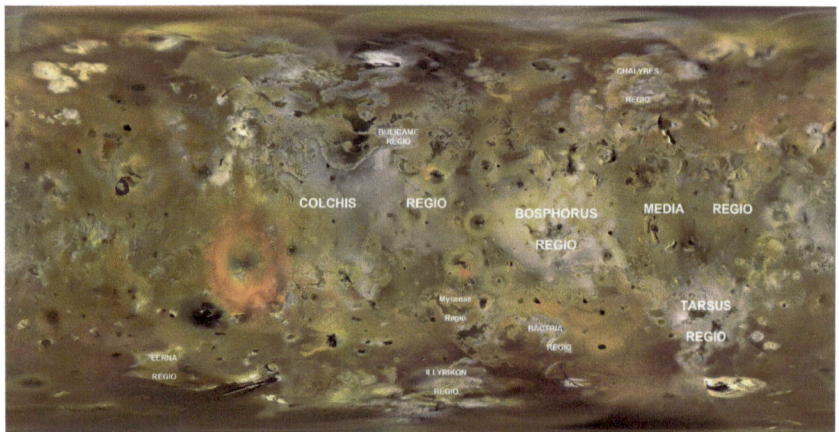

like on Mars 3.866 AU (astronomical units) away. Io is the most

volcanic substance known in our solar system. Io is subjected to

much radiation due to Jupiter causing many things to change the

surface. Io has not just volcanoes of very hot temperatures but

also snow and ice form on the planet. The planet covered with volcanic action, overall has a very cold surface temperature at -130 degrees Celsius. This moon said to be inhabitable by many scientist but it may contain life presently. It is possible that the planet originally around 10 million years ago contained water. Also it could have contained other key but remains of life. Due to the radiation on the moon now all of the water has evaporated and left the moons body. This though does not mean that currently there cant be living species that first originated during that point of 10 million years ago. The species could be under rocks or other formations that can help keep that organism live from the radiation and or any other form of prevention to there survival. It is true that the species must have adapted over the many years to be fine with the current circumstances. This moon has never had a full plan to actually try and reach the surface or investigate further due to the percentage of likeliness compared to other body's, especially there being much more likely ones in the same rotation. Although this is the case this doesn't make it

anymore not likely of containing life in our solar system.

Chapter 4: Titan

Moon makeup

Avg. distance from Sun- 1,426,725,400 km. (Saturn)

Radius- 2,576 km.

Mass- 1.34x10 to the 23rd power km.

Gravity- 1.352 m/s/s

Rotating planet/body- Saturn

Distance coronation to planet- 6th

Size compared to other moons on same orbit- 2nd largest in solar system

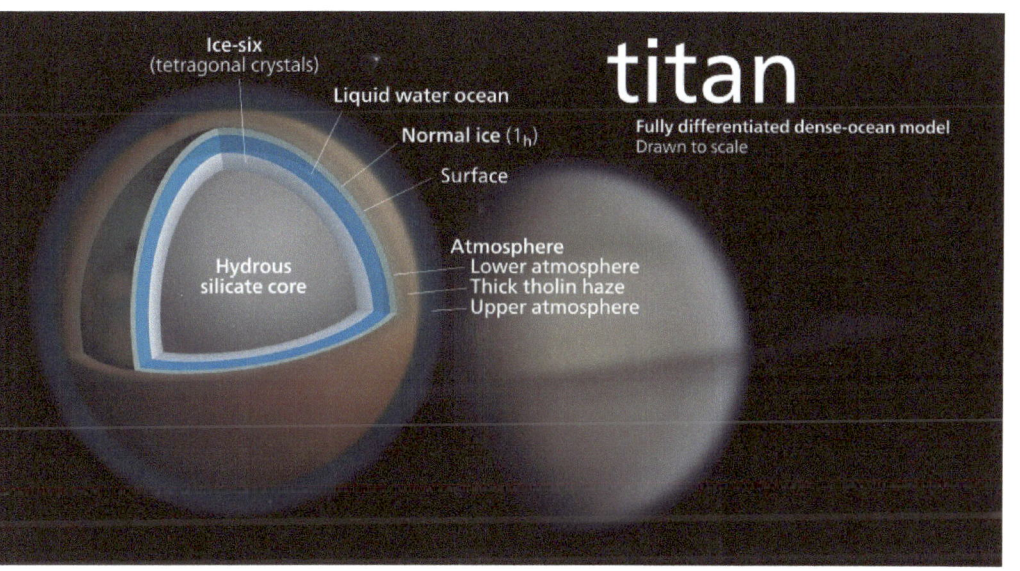

titan

Ice-six
(tetragonal crystals)

Liquid water ocean

Normal ice (1_h)

Surface

Hydrous
silicate core

Atmosphere
Lower atmosphere
Thick tholin haze
Upper atmosphere

Fully differentiated dense-ocean model
Drawn to scale

Titan, the biggest moon of Saturn in fact also the second largest moon in the solar system. It is also very unique that it is the only moon in the solar system to have dense clouds and planet like atmosphere. Titan has been regarded as being very similar to earth many years ago during its earlier time as a planet. This moon is so big it's larger then the planet Mercury With all these planet like features it most be possible that it could contain life and it is. The moon is possibly containing volcanic activity but not yet proven. Titan is predicated in 6 billion years from now after the Earth has been destroyed and the Sun being a red giant, to cause Titans temperatures to increase and may create stable oceans to exist on the moon. If this is the case then Titan will then possible become similar to Earth and allowing conditions that may support life. This in the future can most likely happen based on some computer models done. This isn't the only time life can be on the moon. There could be living organisms on the moon now this is due to the thought that certain organic

chemicals were hovering to high for life to exist.

Chapter 5: Europa

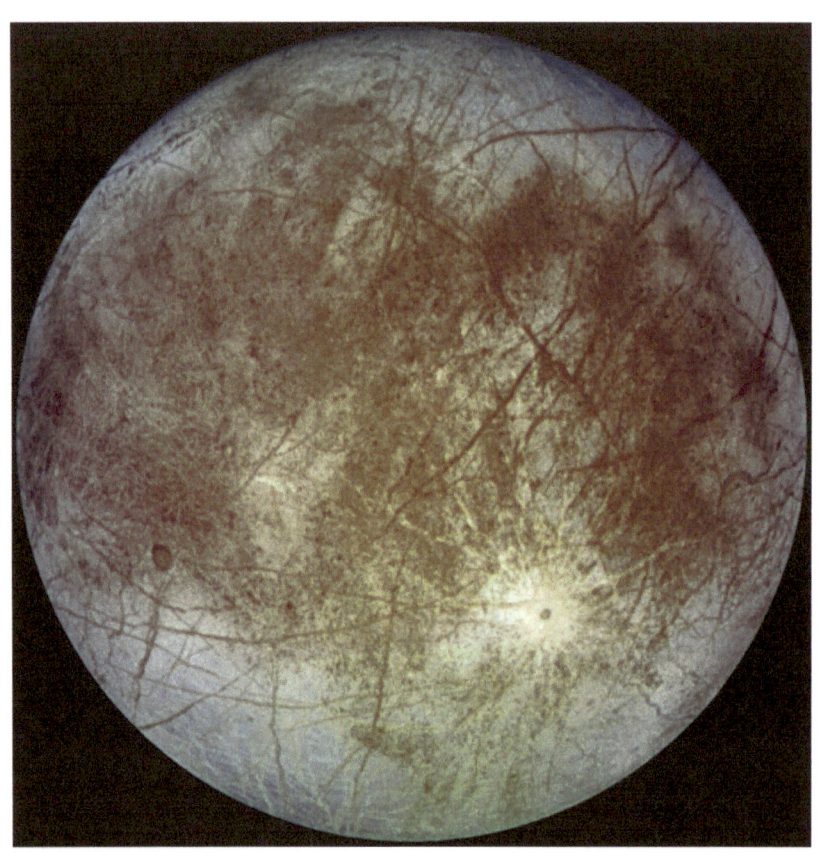

Moon makeup

Avg. distance from Sun- 778,412,020 km.

Radius- 1,561 km.

Mass- 4.8×10 to the 22nd power kg.

Gravity- 1.314 m/s/s

Rotating planet/body- Jupiter

Distance coronation to planet- 2nd closets

Europa, one of the many 67 confirmed moons on the planet

Jupiter. It has been said to may contain life. Europa although

very far from the earth may contain life do to its underground

global ocean. This ocean contains more water then on the whole

planet earth. It is said that we have as humans only explored less

then 5% of the worlds ocean and at least 70% of Earth is the

ocean. Where ever you are in the iceman from the deepest point

all the way to the shallowest there is life. This just shows how much of a possibility there is for life underneath of Europa. Europa underground ocean is protected from Jupiters high radiation. It could very well be a salt ocean like our own that can form life. Jupiter cause this water to heat up and not freeze to be solid ice due to its bending and pulling. This makes it even more likely but it can't be said that there isn't another habitual zone on the moon there are also subsurface lakes possible holding more water then the Great Lakes in the U.S. Also, the Hubble space telescope spotted plumes that were spewing liquid from the ocean underneath. This can be a good starting point for a probe to sample the materials of the water and see if it has the building blocks for life.

Chapter 6: Venus

Planet makeup

Avg. Distance from Sun- 108,208,930 km.

Radius- 6,052 km.

Mass- 4.867x10 to the 24th power kg.

Gravity- 0.904 g.

Length of day- 243 Earth days

Length of Year- 225 Earth days

Surface temperature- 462 degree C

Main Atmospheric Composition- carbon dioxide and Nitrogen

Moons- 0

Venus is a planet not really too familiar with the normal population of people. It is probably due to the fact that there are no current missions and really no significant mission to this planet. Significant meaning in a popular perspective or well known. Venus having no moons and a rocky surface that would kill a person in seconds from many things may actually have life or had life. Although earth and Venus don't really look very similar they are. Underneath the two planets they both have very similar characteristics also, they even are around the same size. Although currently Venus has no ocean or form of water source on its surface, it does have much water vapor in its atmosphere and it can be predicted in past years of the planets existence it could of had a ocean maybe even similar to Earth's. Due to the ultraviolet light of the sun it may have created the water overtime on the planet to break down into atoms of hydrogen and oxygen then break away from the planet into space. There could have been comets that crashed into the planet that gave it water or even life in the past. There are very many possibilities out there but there hasn't been really and big plans to research this in particular. It

seems like the main focus of the earth is on Mars and even farther

planets from our solar system in interstellar space. This does not

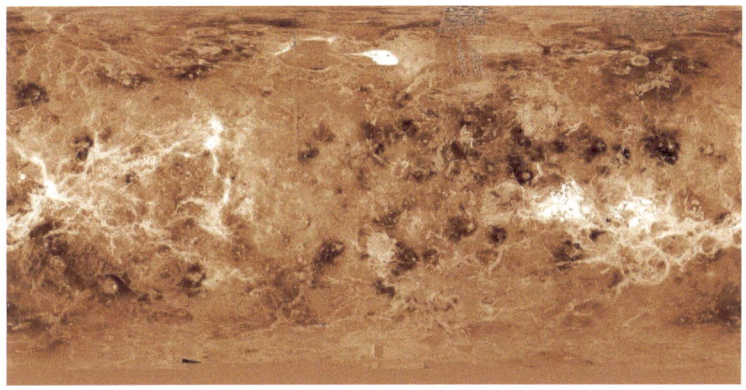

mean that there could not be life currently on the planet. A new

study published has stated that there could be life contained in a

form of hot pressurized carbon dioxide. The theory is that this

substance is almost like how life in water does on Earth. This theory

could also mean that exoplanets in interstellar space which do

contain very high amounts of this substance could contain life and, this theory would change the whole way people search for life in our

solar system and beyond.

Chapter 7: Callisto

Moon makeup

Avg. distance from Sun- 778,412,020 km.

Radius- 2,410 km.

Mass- 1.08×10 to the 23rd power

Gravity- 1.235 m/s/s

Rotating planet/body- Jupiter

Distance coronation to planet- the outer most

Callisto, probably a moon never heard or known commonly. For years scientist have put off the possibility for life on Callisto naming it to be just another moon with a rocky an icy body.

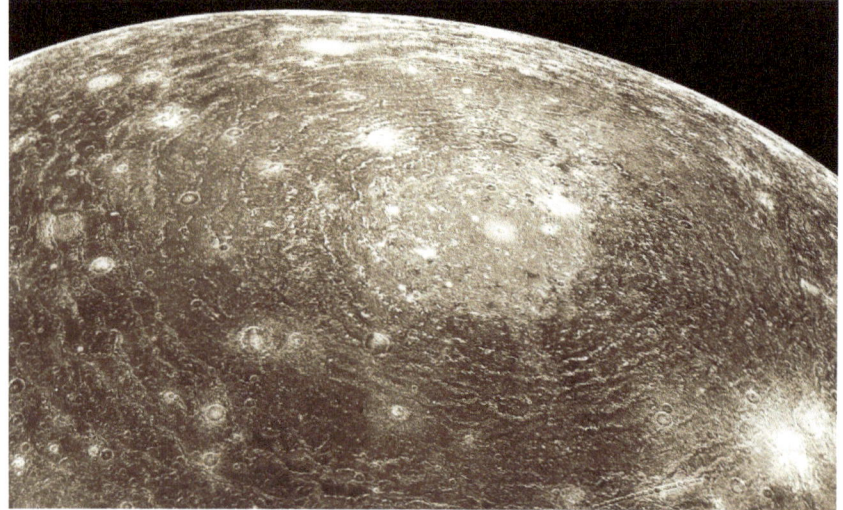

Callisto the outer most moon of Jupiter is quit large having a total of 2,410 km. This is almost the size of the planet Mercury. It's moon though was redetermine when it was stated it could very well have a underwater level or ocean underneath its surface. This is great news for life because on earth where ever there is water there is most likely some form of life. Also Callisto could harbor the right substances vital for life. Callisto is a planet with the most known impact craters in our solar system. This is also great news because some materials or building blocks for life could very well have been planted on that moon from one of or many of the asteroids or space debris that impacted the surface. This moon could very well be and predicted to be billions of years old. This then concluding that it is not known what could have happened to this moon over its long existence to give it vitals for life.

Chapter 8: Ganymede

Moon makeup

Avg. distance from Sun- 778,412,020 km.

Radius- 2,634 km.

Mass- 1.48×10 to the 23rd power km.

Gravity- 1.428 m/s/s

Rotating planet/body- Jupiter

Distance coronation to planet- 2nd farthest out of main moons

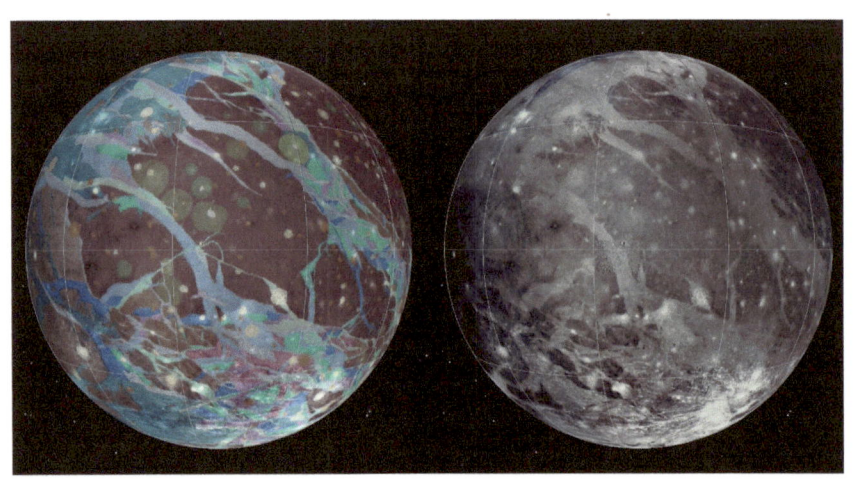

Ganymede a moon almost looking very similar to are own yet very different. Just like another moon surrounding Jupiter being Europa, Ganymede has been predicted to have a global underground ocean. Ganymede is also very special it is one of the only known moons in our solar system to have a magnetic field. This is very important for scientist when looking at the

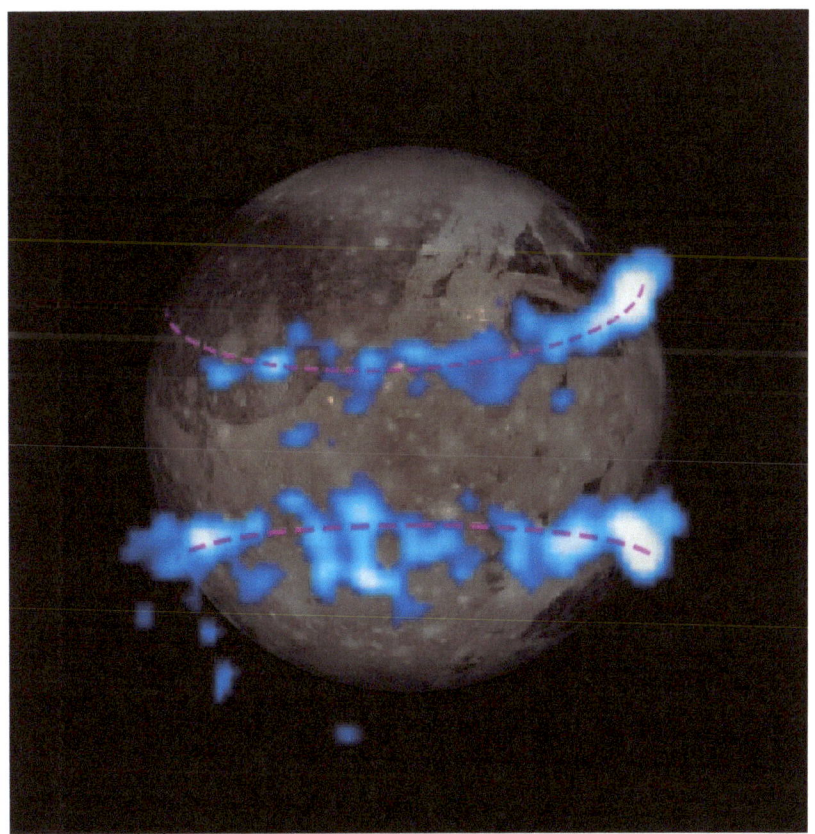

moon having a magnetic field which causes the moon to have

this amazing scenic view similar to our northern lights on Earth.

Though when scientist looked at this view they discovered they

were not as much as they would be without having an ocean

under its surface. After this discovery it was said that it did

indeed have a ocean underneath of it and an ocean with more

water then on earth. Again everywhere water is there's life. The

crust of the surface of Ganymede is mostly ice. Ganymede

ocean is also predicted to be deeper then earths and the ocean

would be covered by total darkness. Well, a way life could have

taken place in Ganymede is through hydrothermal vents under

the surface of the moon. On Earth in the great depths of the

ocean were these things thrive in total darkness there is a whole

ecosystem of life adapted to these climates. The material the hydrothermal vents spew help create life including other things

that most likely are prominent in the deep ocean of Ganymede.

Chapter 9: Enceladus

Moon makeup

Avg. distance from Sun- 1,426,725,400 km.

Radius- 252 km.

Mass- 1.08×10 to the 20th degree kg.

Gravity- 0.113 m/s/s

Rotating planet/body- Saturn

Distance coronation to planet- 2nd innermost

Another one of the most famous for possibly holding a form of life is the icy moon Enceladus. Enceladus was discovered by William Herschel in Aug. 28, 1789. This moon consists of very key features when looking for life in our solar and beyond. Enceladus' very icy surface has a great surprise a subsurface ocean. With any moon orbiting in our solar system discovering a ocean requires more research into its habitability for life. This ocean also is salty. In 2005 the Cassini space probe device read something extraordinary plumes coming from its surface it

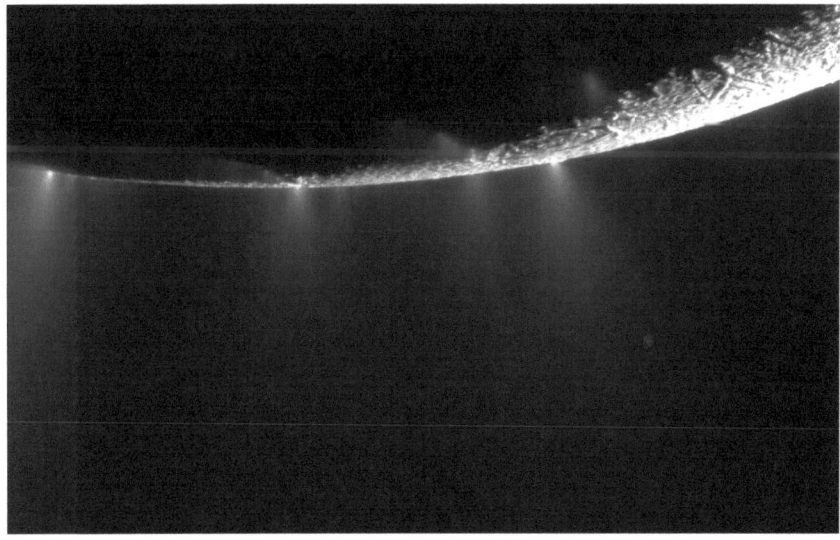

tested these geysers for elements and found water vapor gushing out. Very recently scientist in 2015 confirmed a global ocean. This is very different then many other moons or planets predicted to have a global ocean due to the emended research capability of the moon around Saturn they were able to confirm that there is an ocean which is huge for future missions. It is due basically to the certainty of the water to then propel reasons for future missions. Later called tiger stripes the slices of Enceladus is where the geysers are coming from and these cracks cause the them to break out out of the moons ocean. These geysers help the probe be able to pass through them and discover materials needed for life this is a great brake thorough in proving theories of global oceans of moons and life.

Chapter 10: Conclusion

Life is something now recently discovered to be possibly more

possibly then ever before with the numerous forms of research.

From moons to even planets life can be predicted to live and the

science found proves it. Life is something maybe even

surrounding us. These discoveries are so recent it is not known

when there will be something extraordinary that the Earth

discovers to prove that life does exist beyond are planet maybe

in are solar system. These eight possibility on the surface don't

look like they could hold life but until there was research done

was there actually was a great surprise that there could be and,

that the message should be taken research of anything leads to

things that are unpredictable that the universe is unpredictable.

There are numerous ways life could form that we may not even know possible, life is out there it is only a matter of time till the day we find it.

Work Cited

NASA. NASA. Web. 11 Mar. 2016. <https://www.nasa.gov/ press-release/nasa-confirms-evidence-that-liquid-water-flows- on-today-s-mars>.

"Conditions That Support Life." Conditions That Support Life. Web. 11 Mar. 2016. <http://learn.genetics.utah.edu/content/ astrobiology/conditions/>.

Did Life Begin On Mars?" NPR. NPR. Web. 11 Mar. 2016.

"How Much of the Ocean Have We Explored?" How Much of the Ocean Have We Explored? Web. 11 Mar. 2016.

"Jupiter's Icy Moon Europa: Best Bet for Alien Life?"

Space.com. Web. 11 Mar. 2016.

"Supercritical Carbon Dioxide and Its Potential as a Life-Sustaining Solvent in a Planetary Environment." MDPI. Web. 11 Mar. 2016.

"Was Venus Once a Habitable Planet?" European Space Agency. Web. 11 Mar. 2016.

"Jupiter Moon Ganymede Could Be Teeming with Ocean Life." The Big Science Blog. 2015. Web. 11 Mar. 2016.

"Ocean on Saturn Moon Enceladus May Have Potential Energy Source to Support Life." Space.com. Web.

"Vito Technology Inc. - IPhone, IPad, IPod Touch, IOS, MacOS, Android Educational Apps & Apps for Stargazing and Night Sky Watching." Vito Technology Inc. - IPhone, IPad, IPod Touch, IOS, MacOS, Android Educational Apps & Apps for Stargazing and Night Sky Watching. Web. 24 Mar. 2016.

www.ingramcontent.com/pod-product-compliance
Lightning Source LLC
Chambersburg PA
CBHW040812200526

45159CB00022B/476